RAISE YOUR MATH GRADE

A SHORT, INFORMAL, UBER-STREAMLINED PATH TO SUCCESS IN MATH

ALLISON DILLARD

HAPPY HYPOTENUSE PUBLISHING

The Least You Need to Know Series

Raise Your Math Grade

A Short, Informal, Uber-Streamlined Path
to Success in Math

Allison Dillard

ISBN: 978-1-950720-07-1

AN OPEN EDUCATIONAL RESOURCE

Raise Your Math Grade is an Open Educational Resource. This means that you can share it openly and freely. Pass it along to friends, classmates, teachers, parents, and anyone struggling with math.

Help others follow in your footsteps to success in math.

Anyone can download this book for free at:

raiseyourmathgrade.com.

CONTENTS

Introduction vii

1. Start Here 1
2. What to Do If You're Not Doing Homework 4
3. What to Do If You Don't Have Enough Time for Math 8
4. What to Do If You Study a Ton, But Still Get Bad Math Grades 13
5. How to Succeed Without Asking for Help 17
6. What To Do If Math Is Overwhelming and Stressing You Out 23
7. What to Do If You Just Can't Get Motivated 26
8. What to Do If You Make Careless Errors 30
9. What to Do If You Have Math Anxiety 35
10. What To Do If You're Failing 40
11. Read This Before Your Next Test or Quiz 42

A Note from Allison 45
Book Resources 47
7-Day Raise Your Math Grade Bootcamp 49
Also by Allison Dillard 51
Gratitude 53
About the Author 55

INTRODUCTION

If you're reading this, you've likely got a problem (math) and a deadline (your next quiz or test), and you need someone to help you solve your problem.

You're in luck. I'm a math professor and I've written this ridiculously short book just for you.

In it, I've condensed 20 years of experience teaching and tutoring math into a short, informal, uber-streamlined success path. The entire purpose of this book is to help you raise your math grade as drastically, efficiently, and quickly as humanly possible.

So let's get to it.

-Allison

1. START HERE

There is one super-important thing that students who struggle with math forget.

Or don't understand.

Or maybe were never told in the first place.

You improve with effort.

How do I know that?

Because math is a skill.

Just like learning to play soccer.

Just like learning to play the piano.

Just like learning a new language.

Or how to talk.

Or how to walk.

That's right. Once upon a time, you were a baby who couldn't even walk. You tried it once, fell down, and cried.

But your parents wanted a child who was healthy and enjoyed the freedom that walking provides. They also didn't want to raise a child who gave up so easily that they never even learned to walk, so they encouraged you to pick yourself up and try again.

And again.

And again.

And again.

Do you know what it's called when you work at something over and over and over, despite falling, despite struggling, despite it just being easier to crawl for the rest of your life?

It's called *practice.*

Nothin' special.

Yes, it takes effort.

Yes, it can be boring.

And tedious.

And hard.

But that's how you improve when you want to develop a skill.

And math is a skill.

So take whatever nonsense you've embraced about not being a "math person" and shove it—wherever you shove ideas that are counterproductive to growing into a fully functioning grown-up.

When it comes to ***math***,
you improve with effort.

When it comes to ***any skill***,
you improve with effort.

Practice.

Put in the effort.

And you will raise your math grade.

2. WHAT TO DO IF YOU'RE NOT DOING HOMEWORK

You improve with effort, but not all effort is created equal, and there's no point in putting effort into things that won't raise your math grade.

So where should you direct your efforts?

Which effort is the least likely to waste your time?

Which effort will help you the most to prepare for your math test?

The answer to these questions is:

Your homework.

Do your homework.

Homework is not optional.

It's not an unrealistic expectation.

It is not a pointless task assigned by your teacher to be mean.

Or stress you out.

Or make you feel stupid.

Your teacher assigns homework because they know that math is a skill and you must practice to develop the skill.

Homework provides that practice, so...

Do your homework.

I understand, however, that this is easier said than done.

After all, if you're not getting through your homework, there are likely good reasons for that.

Don't skip over the reasons. What are they?

Acknowledge them.

Write them down.

And brainstorm ways to overcome them.

Then change up how you do math homework until you find a way to actually get through it.

For example, suppose the primary reason that you don't get through homework is that it takes too long. It's supposed to take an hour, but you're an hour in and not even a quarter of the way through.

What should you do?

Brainstorm different ways to overcome the problem.

For example, if I were struggling to get my homework done, here are a few possible solutions I'd brainstorm:

1. Text a friend for help.
2. Put my phone in the other room while I do my homework, so it doesn't distract me.
3. Do math homework first, instead of last, at the end of the day when I'm mentally exhausted.
4. Email my teacher for help on a specific problem.
5. Have a snack before homework, so I have the energy to think.
6. Listen to different music that isn't distracting.
7. Go to my school's tutoring center or math center for help.
8. Give myself a time limit. If a question takes more than 5 minutes, move on to the next.
9. Do homework with a friend who actually finishes their homework.

There you go. One problem (homework taking too long) with nine potential solutions.

When you're brainstorming solutions to a problem, nine is a good number. Nine gets you more ideas and gives you more options. Nine forces you to be proactive and think deeper about how you're approaching your problem. It gets you to be creative and think outside the box.

Whatever reasons are preventing you from doing homework, brainstorm nine changes you can make to overcome those reasons.

You aren't powerless.

Be proactive.

Change what you're doing.

And keep changing it up until you've figured out how to get your homework done.

Additional Resource: If you need help brainstorming solutions to your problems, I have a free downloadable "Rule of 9 Worksheet" to walk you through it at raiseyourmathgrade.com/resources.

3. WHAT TO DO IF YOU DON'T HAVE ENOUGH TIME FOR MATH

But what if you're overscheduled and just don't have time for homework or studying math?

You're not going to like my answer here.

And I know I'll sound like a jerk for saying it...

But if you're overscheduled and don't have time for your math homework...

Then...

Make time for math.

Here's the thing about time.

Everybody gets twenty-four hours each day.

Everybody.

You.

Me.

The kid getting the highest grade in your math class.

The kid getting the lowest grade in your math class.

And everybody in between.

24 hours.

That's it.

So if we all have the same number of hours in the day, then this really is not a time issue.

It's either an efficiency issue—if homework takes too long, see the previous chapter—or a prioritization issue.

Let's get your priorities straight.

You should have only one.

Or two.

Or three priorities, at the absolute most.

Do this now: List out your top three priorities and schedule them into your day—the earlier in your day, the better.

Your list might look like this:

1. Job.
2. Math.
3. Another class.

Or it might look like this:

1. Math.
2. Sport.
3. Job.

But if it looks like this:

1. Job.
2. Class #1.
3. Class #2.
4. Class #3.
5. Math.
6. Sport.
7. Extracurricular activity.
8. Hanging out with my boyfriend.
9. Walking my dog.
10. Babysitting my little sister.
11. Maintaining a presence on Instagram.
12. Making dinner.
13. Playing video games.
14. Watching TV.

Then, no wonder you're not raising your math grade.

A priority, by definition, is a thing that is more important than other things.

If you have 14 "priorities" on your list, then you don't have a list of priorities. You just have a list of a bunch of stuff that you do.

Whittle this list down to (ideally) one, (realistically) two, or (at the absolute max) three priorities.

Then, ask yourself the following questions:

If you have to choose between studying for a class that's an elective and studying for math, which *should* you sacrifice?

If you have to choose between going to your piano lesson and studying for math, which *should* you sacrifice?

If you have to choose between watching TV and studying math, which *should* you sacrifice?

If raising your math grade really is a priority, then time for it should be scheduled.

Scheduled as early as possible in the day.

Scheduled as often as needed to raise your grade.

Does this situation suck?

Yes.

But making progress towards your priorities requires sacrificing time for other things.

And if the situation is bad, you'll have to sacrifice something that's awesome and important in its own right if math is a higher priority.

I skipped a soccer game once to do math homework.

I remember it so clearly.

That's how awful it was.

My soccer game was the thing I was looking forward to the most that week.

Every week, Thursday night, it was the thing I most looked forward to.

Biggest sacrifice ever.

And I let down my team. (They had to play without a full team and lost the game.)

So much guilt.

And I hated this math class.

It was hands down my least favorite math class of all time.

But it was a requirement for my Master's and I couldn't drop it.

I knew I had ZERO chance of getting an extension.

I knew my professor would fail me in a heartbeat if he had the chance.

And I knew that, if I spent the evening playing soccer, I had ZERO chance of finishing the assignment... I was that in over my head.

I didn't even know if I COULD finish the assignment.

But I had to try.

I had to make *trying* a priority.

This class was that crucial to the future I envisioned for myself, so when push came to shove it took priority over the thing I enjoyed most.

In a nutshell, if you don't have time for math:

1. Have no more than 3 priorities.
2. Get your priorities done first.
3. Even if it means making sacrifices.

Do math first. Future You (you know, You, but at some point in the future) will thank you for it.

Additional Resource: If you need more help, check out my free downloadable "Priorities and Time Management Worksheet" at raiseyourmathgrade.com/resources.

4. WHAT TO DO IF YOU STUDY A TON, BUT STILL GET BAD MATH GRADES

Holy moly!

If there's one thing I learned from all my years as a tutor, it's that some of you guys and girls are bad at studying. You can literally review your homework, notes, and textbook for hours—literally hours—and still not be prepared for your test.

If reviewing your notes, homework, and textbook does not sufficiently prepare you for your test, it does not count as studying.

It counts as wasting time on your phone, while you pretend to study.

It counts as reading about math, when you should be preparing for your math test.

It counts as wasting your time and making yourself miserable, while you could be doing anything else in the world, like having fun or—perish the thought—*actually studying* for your math test.

So if "reviewing" is not getting you the grades you want, here is how to study in a way that will better prepare you for test day:

Quiz yourself.

How do you do that?

1. Guess at what questions will be on your test.
2. Set a timer and go. Solve them in *a test setting*, meaning don't use notes, books, Google, or anything else that is not permitted during your test.
3. Review your test. Correct it. Learn from mistakes, so that you don't make them again on the actual test.
4. Repeat, as needed.

To raise your math grade, stop reviewing and, instead, quiz yourself.

I know.

I can feel your eyes glazing over.

I'm losing you.

So let's mix it up and talk about swimming.

Suppose you're a swimmer and you have a swim meet coming up.

What is the best way to prepare for your swim meet?

One way to prepare is to sit around in your pajamas and read about swimming in books and notes.

Perhaps all this "reviewing" will lead you to an "Aha!"

moment where you say, "Yes! That backstroke techniques makes sense to me! That is what I'll do during the race!"

Noooooooooo!

That's NOT how you prepare for a swim meet!

You jump in the water and SWIM to prepare for a swim meet. You swim in the same conditions, with the same swim style, the same distance.

No matter how many hours you spend reading and reviewing about swimming, if you insist on never jumping into the pool before a swim meet, you'll never show up prepared.

Yet, this is exactly the strategy that struggling math students use to prepare for their math tests.

They review and read and look over problems.

When in fact, they need to be quizzing themselves, solving math problems in a test setting.

So take a guess.

Pick out a couple homework questions or problems from your book.

Write them down on a blank piece of paper.

Set a timer.

And see if you can solve them in a test setting.

If you get it wrong, learn from your mistakes, so you don't make the same mistakes on your math test.

Repeat, as needed.

Additional Resource: You can get a free downloadable "Practice Test Template" that you can use to take your first practice test at raiseyourmathgrade.com/resources.

5. HOW TO SUCCEED WITHOUT ASKING FOR HELP

Asking for help is lame.

And awkward.

And a sign of weakness.

You wouldn't know what to ask anyways.

Or who to ask.

Or if it'll even be helpful.

In fact, it might make you feel dumber.

Or more confused.

Or verify that you know way less than you let on.

So best just to skip asking for help and figure it out on your own.

Cool.

I'm glad we cleared that up.

Let's detour a few minutes to talk about soccer instead.

I'm finding myself in a pickle and could use your advice.

You see, I recently joined a soccer team and this team is waaaay better than I am. I'm not sure how I got myself into this situation, but everyone's skill level is a notch above where I'm at.

And I desperately want to avoid failing, being cut from the team.

It's my #1 priority, in fact.

I want to improve my soccer game as drastically, efficiently, and quickly as possible.

What do *you* think I should do?

Figure it out on your own.

Figure it out on my own? I knew you'd get it! That's why I asked you, of all people, for advice.

After all, that's the strong thing to do.

The independent thing to do.

The self-reliant thing to do.

So that's what I did!

I went out to the field and practiced a whole bunch on my own.

Then, I did a Google search, watched some YouTube videos, and practiced the skills I thought I needed.

Awesome. Problem solved.

Except.

Now, it's a few weeks later and...

I'm still the worst player on the team.

I was able to improve some skills on my own, but I struggled with others. Some things I understand when I practice them on my own, but then doing it in the game is sooo much harder. I just forget. And, worst of all, the coach made a comment that I'm kicking the ball the wrong way-- I'm trying not to think about the fact that I might have learned to do certain things the wrong way.

Something isn't working, but, if I'm being honest, I either don't know what it is or I don't know how to fix it.

My #1 priority is still to get good enough I stay on the team, but... How? What should I do?

Ask for help.

No! The reason that I'm asking you, YOU, of all people, is because I don't want to ask for help.

But maybe your coach--

No buts! I understand that my team has a coach, a person who has decided that of all the jobs in the world, the one to which they wanted to dedicate their lives was teaching students to play soccer. I get that!

I don't want to ask my coach for help.

I wouldn't even know how, or when, or where to ask my coach for help. It's not an option.

Doesn't your coach have office hours?

Yes, of course, my coach holds office hours. He's always going on about how the sole purpose of office hours is to help soccer players who need some extra help and we should come see him.

So?

So, I'm avoiding office hours, LIKE THE PLAGUE!

But--

No buts! We established at the beginning that asking for help is lame.

And awkward.

And it's gonna make me feel stupid.

And it might not help anyways.

So I'm just going to pretend to be invisible and hope my coach either doesn't notice me or forgets who I am until I improve.

Maybe ask a teammate for help?

I did that, but Rick doesn't know how to kick a soccer ball either and I think we just confused each other more.

Maybe ask a teammate who knows what they're doing?

No, they're totally unapproachable.

And it would be awkward.

They'd probably just make me feel stupid.

And it probably wouldn't help anyway.

Maybe go to the training center for help?

Ugh, my coach is always going on and on about the training center. It's like the math department's tutoring center, but for soccer. I get it! That's where "the best soccer players" in the league, who have been selected "specifically for their teaching abilities," hang out, "with the sole purpose of helping struggling soccer players" succeed.

Of course, I'm not going there! What kind of advice are you giving me?

So, you're saying that you'll reach your goal faster if you avoid this training center—designed and funded specifically to help you —like the plague?

No, I'm saying I want to succeed despite that.

I don't know what to tell you. You're not gonna improve drastically, efficiently, and quickly if you avoid everyone who can help you.

I'm just too shy.

It'd make me uncomfortable.

And it'll be awkward.

It sounds like you might get cut from the team then.

I know. It sucks. It's too bad there's nothing I could do.

The end.

What's the moral of that story?

Ultimately, to be successful, you must get used to doing the things that make you uncomfortable.

So, this is my challenge to you:

Move out of your comfort zone.

Ask your math teacher for help.

Go to your teacher's office hours.

Ask the best math students, not other failing or struggling math students, for advice.

Ask for help at your school's math center or tutoring center.

Do all these things, despite how uncomfortable it makes you feel, because that is how you will drastically, efficiently, and quickly raise your math grade.

From my experience, this is obvious to everyone *except* students who are struggling in math.

Additional Resource: If you don't know what to say or what to write in an email to your teacher (or anyone you're asking for help), don't let that hold you back. You're welcome to use my "How to Ask Your Teacher for Help" Email Templates at raiseyourmathgrade.com/resources. Copy, paste, and get the help you need.

6. WHAT TO DO IF MATH IS OVERWHELMING AND STRESSING YOU OUT

To raise your math grade, you may need to catch up on all of your homework, replace "reviewing" with quizzing yourself, and reach out for help.

You may also need to strengthen your math foundations, learn new test-taking skills, and crush the self-defeating notion of "not being a math person."

Following through with these changes will also require you to up your motivation, discipline, and grit.

It's a lot to tackle at once. There are many components that must fall into place. When you have so much to take on in such a short period of time, it's easy to get overwhelmed and give up.

So what should you do when it starts to get overwhelming and stressful? When the idea of giving up or not getting started is tempting you?

Create a game plan.

I like game plans better than study plans.

While study plans are fixed, rigid, and rarely work as well as you hope, game plans are fluid, ever-changing, and far more effective.

A game plan focuses on what you are doing and what *changes* you are making to raise your math grade.

As your situation, for better or worse, changes, so does your game plan.

But what should your game plan include?

In its most simple form, a game plan answers one question:

What will you do to achieve your goal?

A game plan is all about what you do.

The work you put in.

The part that you control.

The changes that you make.

Focus on change.

If you're not doing your homework, do your homework.

If you bombed your last test, change how you study for the next one.

If that last person you went to for help wasn't helpful, then go to someone else for help.

If you're not putting in the time, put in the time.

If your foundations are weak, but you've been carrying on despite that—go back and strengthen your foundations.

Your game plan should focus on change.

Albert Einstein once said, "Insanity is doing the same thing over and over again and expecting different results."

Don't go insane.

Create a game plan and make the changes you need to drastically, efficiently, and quickly raise your math grade.

Additional Resource: To help get you started, I've created a "7-Day Raise Your Math Grade Game Plan" for you to print and fill out. You can download it for free at raiseyourmathgrade.com/resources.

7. WHAT TO DO IF YOU JUST CAN'T GET MOTIVATED

You know that you should do your homework, but you don't do it.

You know you should study for that test ahead of time, but you don't.

You know you should ask your teacher questions, but you don't.

The problem isn't that you can't figure out math or that you aren't a "math person." The problem is that you don't have the motivation to do what you need to do.

If you don't think motivation can really be a problem, meet Jared.

Jared worked as a valet. His mom was making him go to community college. She called me and asked me to tutor Jared the night before his Statistics final exam.

I will tell you right now, this kid was not motivated. He never did his homework (it was optional), he'd never gone to class (also optional), and he was hoping to pass on pure luck. As the wise Stats tutor that I was, I left him with a list of exactly what he needed to review for the remainder of the night. All he had to do was cram for one night and, because his final was worth so much of his grade, he had a shot at passing.

Do you think he did the work after I left?

Do you think he passed?

Heck no!

After I left, he didn't look at his book once.

He *kinda* wanted to pass.

But not enough to do his homework.

Not enough to show up to class.

Not even enough to study the night before his final exam.

Of course, he failed.

Here's the interesting part though...

Two years later, I got another call from Jared. Not his mom this time, but Jared himself. He was still working as a valet and wanted to schedule tutoring sessions with me for Statistics. This time, we started in the first week of class, and he intended to study at both the math center and with me every week of the semester. On top of that, he was paying for school and our $100+ per hour tutoring himself. *Why*? Because there was no way he was going to work one more year as a valet. He was sick of it, and school was his way out. The same kid who couldn't bring himself to study the night before his final now had motivation.

He passed Stats this time, and the only thing that changed was his motivation.

If you lack motivation, nothing else in this book will help you.

No parent or teacher or tutor can help you.

No tip or hack or study strategy can help you.

So it's best to fix the motivation problem *now*.

How do you get motivated?

Find a goal.

Then, write it down.

First, you need an important goal. A goal that you *have* to achieve, that's crucial to you becoming the person you want to be. A goal that's worth making sacrifices.

Second, you need a big goal. A goal so big that it intimidates you because you know that in order to achieve it you'll have to grow into a stronger, smarter, grittier version of yourself.

Third, you must *understand why math is essential to your goal.* If you can't connect math to your goals, you'll cave to the temptation to scroll through Instagram, play video games, or text for five hours straight.

I'm not kidding.

1. Big goals.
2. Important goals.
3. Goals that can only be achieved if you raise your math grade.

Don't underestimate the power of a goal being *important* to you.

The first time I worked with Jared, his goal was to pass Statistics, so he could transfer to a 4-year college. *Why?* To keep his mom from nagging him.

The second time I worked with Jared, his goal was to pass Statistics, so he could transfer to a 4-year college. *Why?* Because he couldn't bare to work one more day as a valet. He hated it with his entire being and would do anything to be qualified for something better. College was his only way out of his current life and passing Statistics was a requirement to transfer.

The first time, he couldn't care less about his goal. The second time, who he wanted to be, the daily life he wanted to live, and the entire future he envisioned for himself were all at stake.

You only get one life. Don't be intimidated and make it small.

Dream big. Learn some math and make your life everything you want it to be.

Additional Resource: If you struggle with motivation, check out the quick routine I use every day to get and stay motivated. You can download it for free at raiseyourmathgrade.-com/resources.

8. WHAT TO DO IF YOU MAKE CARELESS ERRORS

You know you studied and you understand the material, but you just make "careless errors."

It happens by accident, even though you understand the material.

And there doesn't seem to be anything you can do because you prepared.

You did your job studying.

It's the careless errors that get you.

It's out of your control.

And, if there's no way to control it, then there is no way to fix it.

It's frustrating, but, in a way, it lets you off the hook.

You don't even have to try because, in this particular instance, effort is futile.

My friend, what do you think I'm going to advise you here?

That careless errors are, in fact, out of your control?

That you do, in fact, know the material as well as you think you do?

Of course not.

Blaming your struggles in math to careless errors is much like blaming your struggles in math on "not being a math person." Both give you permission not to try.

So just as some people need to crush the crazy idea they're "not a math person," you also need to crush the idea that you're struggling because of "careless errors."

Errors are never careless.

Something always causes them.

And you need to get to the bottom of what is causing your errors.

Here's the process I use to help students identify and eliminate errors:

1. **Figure out what errors you're making.** You can do this with the *Know It/Don't Know It* List, which is downloadable for free at raiseyourmathgrade.com. Every single time you make a mistake, write down *what* the mistake was. Trust me, by writing them down every time, it quickly becomes obvious which mistakes you make over and over and over again.
2. **Learn from your mistakes.** Review your mistakes with a teacher, tutor, parent, or friend—someone

who can help you understand what the mistake is and why you made it. Often times, these errors are caused by weak math foundations. (Having weak math foundations means that you never learned the basic skills or prerequisite material *well enough to use it automatically and without error in more complicated math questions*.)

3. **Practice.** Reduce the chances of making these careless errors through repeated practice. Practice the foundations you need to strengthen. Take quizzes with just the questions you make errors on and practice looking for errors.
4. **Check your work.** At the end of your next quiz or test, don't check everything. Look specifically for the types of errors you've made in the past and see if you can catch any.

I get that the process is boring.

You track your errors.

You make sure you understand them.

You practice not making them.

You track and practice, repeatedly and obsessively.

And, on top of that, the process is humbling.

Brian, another student I tutored, was failing Statistics until I was hired to help him out.

When we tallied up his careless errors, we found errors with multiplication, division, and simplifying fractions.

Oh, you should've seen the look on his face when I told him that he didn't know *multiplication* well enough.

Clearly, he knew how to multiply!

Six times seven was forty-two!

To say that he didn't know such basic math was insulting.

I wasn't saying that he didn't know it at all.

I was, however, saying that he didn't know it *well enough to use it automatically and without error in more complicated math questions*.

So he made flashcards for basic multiplication and division and reviewed them daily.

He also printed out and completed worksheets on simplifying fractions.

He spent extra time working on and checking problems that required multiplication, division, and simplifying fractions.

At the end of his midterm, he went back and checked his multiplication, division, and steps for simplifying fractions.

Did he pass?

Of course.

Why?

Because, as humbling as it was, he took his errors seriously and did the hard, tedious work necessary to stop making them.

Your errors are not careless.

Tackle them head on.

And eliminate them. *Now*.

Additional Resource: You can find my "Know It/Don't Know It List" and other free resources that can help you to quickly strengthen your math foundations at raiseyourmathgrade.com/resources.

9. WHAT TO DO IF YOU HAVE MATH ANXIETY

You're "not a math person" in the extreme.

Maybe homework makes you nauseous.

Maybe you blank out on quizzes.

Maybe you even hyperventilate, think you're having a heart attack, and throw up as your teacher passes out tests.

Ugh!

Math anxiety is the worst.

I'm getting stressed out just writing about it!

What do you do if you have math anxiety?

As in previous chapters, you may not like my answer.

So roll up your sleeves,

Take a deep breath,

And try to listen with an open mind.

What I'm about to tell you is likely different (albeit more

effective) than what most sympathetic teachers, parents or friends are telling you.

If you have math anxiety, you need to...

Be tougher on yourself.

Like I said, keep an open mind.

Of the many wise things Eleanor Roosevelt has said, my favorite is, "Do the thing you think you cannot do." If showing up to your math test confidently and calmly is the thing you think you cannot do, I challenge you to *become* the kind of person who can.

And, in order to do that, you must be tougher on yourself.

Math doesn't get easier.

You, my friend, must get stronger.

But how?

Read the following story about Priscilla.

Before seeing Priscilla succeed, I used to baby students with math anxiety. I would sympathize with them and hold their hand every step of the way.

My experience working with Priscilla drastically changed how I view and work with students who have math anxiety.

Now, I take a tougher approach.

And I think you should be tougher on yourself also.

But first...

Meet Priscilla.

The first student I ever tutored who had *severe* math anxiety, which included blanking out, having panic attacks, and hyperventilating during exams.

She had failed Precalculus twice and dropped it a third time —so her doubts were not unfounded.

She had every reason to think she'd fail again.

So what did we do?

We took the "failure is not an option" approach.

As tough as it was, she did every single thing I asked her to do.

She tried every single strategy or technique I asked her to try.

She completed every single extra assignment I assigned to her.

Tough was the understatement of the century, but it's what got her through the following:

First, we followed all the instructions I outlined for you earlier to complete 100% of her homework.

Second, we followed the instructions I gave you in Chapter 4 to study by quizzing yourself instead of "reviewing."

Then, only then, once we were certain that her anxiety did not stem from lack of preparation, did we tackle her test anxiety.

We talked about the origins of her anxiety. Yup, Psychology 101. What experiences led you to believe you're "bad at math?"

We made lists of questions she was afraid to see on the test and took extra quizzes on those questions.

I had her write down every negative belief she repeatedly told herself, like "I can't do it," and actively worked with her to replace them with more positive beliefs, like "I can't do it *yet*."

I had her visualize doing math problems, especially complicated anxiety-inducing problems with multiple steps, on a test. (Visualization is a technique commonly used by top students, Olympic athletes, and others.)

I helped her develop a study routine that focused on answering questions in a test setting: I created formal-looking, typed-up practice tests and she took them at the library (a location that felt more formal than at home and reminded her of her classroom).

I had her create a pre-test routine that included meditation, writing about her fears and anxiety, and warming up with a few easy questions.

I had her develop a test routine that included writing down any formulas she might forget as soon as she got the test, answering easy questions first, circling hard problems and coming back to them later, and closing her eyes and taking deep breaths when she started to feel overwhelmed.

And it worked.

You'll be happy to know that, after failing twice and withdrawing a third time, Priscilla went on to pass Precalculus with a B+ and then pass two more semesters of Calculus with A's.

Toughen up.

Experiment.

And never give up until you've crushed your math anxiety.

If Priscilla did it, so can you.

Additional Resource: Be sure to check out my extra resources for crushing math and test anxiety at raiseyourmathgrade.com/resources.

10. WHAT TO DO IF YOU'RE FAILING

If you're failing, you need to go into emergency mode.

Ground yourself until your next quiz or test.

Temporarily cut out all video games.

All television.

All social media.

All fun.

All extracurricular activities.

If you've got sick days or vacation saved up and you can get away with it, I'd even skip work.

Then, do the following:

1. **Create a game plan.** Create a game plan to raise your grade on your next quiz or test.
2. **Reach out for math help.** Then, schedule multiple follow-up meetings between now and your test to keep you on track.

3. **Reach out for emotional support.** Who is a friend or parent who encourages you when you're down? Share your earnest desire to succeed in math with them. Share your game plan with them. Ask them to be your accountability partner and schedule multiple follow-ups with them, so they can help you to persevere when you feel overwhelmed and want to give up.
4. **Schedule extra time to spend on math.** Make appointments with yourself. Put it in your calendar. As much time as possible. (You can always cut back later.) Go all in. You'd be shocked at how much you can learn in a short period of time if you do.
5. **Accept responsibility.** There is no fairy godmother who will appear, wave a magic wand, and make math easy. Realize that your success in math is on you. Your ability to crush math relies on the actions you take, the changes you make, and the work you put in.

Reach out for help.

Take responsibility for your own success.

Study more than before, differently than before, and better than before.

Try and you just might raise your math grade more drastically, efficiently, and quickly than you thought possible.

Additional Resource: Fill out my free downloadable "7-Day Raise Your Math Grade Game Plan" at raiseyourmathgrade.-com/resources, then follow it for a week. You'll be amazed at how drastically, efficiently, and quickly you can raise your math grade then you go all in.

11. READ THIS BEFORE YOUR NEXT TEST OR QUIZ

Nothing will raise your grade more than a thorough understanding of the material.

That said, you should use every advantage in your arsenal to succeed, and these "hacks" should most definitely be in your arsenal.

Students who are already succeeding in math call these strategies common sense. Students who struggle in math call these strategies game-changers:

1. **Review anything you might forget *right before* your math test, then write it down as soon as you get your test.** Even before you read the directions.
2. **If you have math anxiety, do a couple of warm-up problems (easy or medium questions that you know how to solve).** Just like you'd warm up your body before a track meet, warming up your brain a little before test time will get you ready to go.
3. **Skim through the test; glance over the whole test so you know what you're up against.** If some

problems are worth more points than others, take note. These should get more time. If some questions are harder than others, take note. These also warrant more time.

4. **Read the directions!** I cannot articulate how bonkers it makes me when students lose points just because they forgot to read the directions!
5. **Show your work.** This allows teachers to give you partial credit if your final answer is wrong. It also reduces mistakes and helps you catch errors.

The highest grade jump I've ever seen was Aika's. She went from 43% on her first exam to 91% on her second, from an F to an A! It turned out that, on her first exam, she knew more than her grade suggested, but she had panicked and blanked out on the formulas. Without the formulas, she couldn't even get started on most questions, so she failed. On the second exam, she wrote the formulas down as soon as she got the test (Strategy #1 above) and she was able to figure out the rest from there. That was all Aika needed to ace the test.

I believe that you can be like Aika.

I believe you can ace your next math test.

I believe you can raise your math grade—quickly, efficiently, and drastically.

I believe that whatever obstacles you face, you can absolutely overcome them to succeed in math.

I believe this because I've seen it happen over and over and over again.

You've got this. Now, go get to work.

A NOTE FROM ALLISON

Dear reader,

My name is Allison Dillard. I am an adjunct math professor at Irvine Valley College and a private math tutor. For over 20 years, I've taught and tutored students in math, helping them, day in and day out, figure out how to raise their math grade.

I had some health issues a few years back, so I took a year off of teaching and tutoring. After surgery and chemotherapy, I recovered quickly and this unfortunate year off turned out to be a blessing in disguise.

It blessed me with time to reflect on the many reasons that students struggle with math and what strategies best help them to succeed.

It blessed me with the motivation to write *Crush Math Now* and now *Raise Your Math Grade*, putting pen to paper for over a year to figure out how to best share with you the advice I've repeatedly given students over the years.

And it blessed me with the courage to share this book with you, so that I might empower you and other students beyond my own classroom to succeed in math.

Raise Your Math Grade is an Open Educational Resource. This means that you can share it openly and freely. Pass it on to teachers, advisors, and parents. Share it with students, classes, and departments.

Anyone can download *Raise Your Math Grad*e for free at raiseyourmathgrade.com.

To your success!

- Allison

BOOK RESOURCES

As mentioned throughout the book, I've created some free downloadable worksheets to help you get the most of out Raise Your Math Grade:

- A Rule of 9 Worksheet because a common obstacle to success is thinking your problems have no solutions.
- A Priorities and Time Management Worksheet because a common obstacle to success is *not* making math a priority.
- A Practice Test Template because a common obstacle to success is *never* getting around to creating your first practice test.
- An "Ask Your Teacher for Help" Email Template because a common obstacle to success is never getting around to reaching out for help.
- A Game Plan to Raise Your Math Grade because it's easier to simply fill out a worksheet than create a game plan from scratch.
- My Morning Routine because a simple 5-minute

routine *can* have a huge impact on your motivation, focus, and confidence.

- A Know It/Don't Know It List because it drastically streamlines what you need to study for exams, which saves you time.
- Extra Resources on Math Anxiety because anxiety is not an easy obstacle to overcome.

You can download all these resources for free at raiseyourmathgrade.com/resources.

7-DAY RAISE YOUR MATH GRADE BOOTCAMP

For more help raising your math grade as drastically, efficiently, and quickly as possible, sign up for Allison's 7-Day Crush Math Now Bootcamp at allisonlovesmath.com/bootcamp

Here, Allison will:

- Help you create a one-week game plan to raise your math grade,
- Teach you simple strategies to ensure you actually do the things you need to do,
- And help you daily to adjust your game plan, as needed.

One goal. One week.

Let's raise your math grade!

ALSO BY ALLISON DILLARD

Allison Dillard's more comprehensive book, *Crush Math Now,* takes a deeper dive into everything you need to do to raise your math grade.

Get your copy today on Amazon!

GRATITUDE

I could not have completed Raise Your Math Grade without the help and support of some amazing people:

My husband and best friend, Joe.

My children, Zoey, Abby and Matt.

My family, Mom, Dad and Christy.

My book launch team, Catherine, Beth, Jamie, Alicia, Monica, Erika, Amy V., Clarissa, Tracy, Kacey, and Colleen.

My book team, Susan Dimmock and Jerald Anino.

The many amazing math professors at Irvine Valley College, including Brent Monte, Joel Sheldon, JoAnne Noyes, Rich Zucker, Ilknur Erbas White, Lan Pham and Carlo Chan.

And, as special thanks to, Megan Haskell, Honoree Corder, Pete Vargas, Pat Quinn, and Hope Zvara.

ABOUT THE AUTHOR

Allison Dillard is the author of two books, *Crush Math Now* and *Raise Your Math Grade.*

She is an adjunct math professor at Irvine Valley College, has over 20 years of experience tutoring and teaching math, and has previously trained new SAT tutors for Kaplan Test Prep.

She has an M.S. in Mathematics from Claremont Graduate University, where she was the recipient of fellowships from the National Science Foundation, Boeing, and Los Alamos National Lab. She has a B.A. in English and a B.A. in Mathematics from Scripps College, where she received the Reed Institute's award for "Excellence in Problem Analysis and Problem Solving in Decision Science."

When she's not teaching or writing, you'll find her playing boardgames with her awesome husband, coaching her kids' soccer teams, and dreaming up new ways to convince the world that math is awesome.

Her love for math is legendary.

You can connect with her at allisonlovesmath.com.

Made in the USA
Monee, IL
06 November 2026

67974632R00038